Advance Praise for Lawrence Millman's Goodbye, Ice

"*I imagine future archaeologists finding wind-chiseled stones in an Inuit graveyard. On each stone is carved a poem from Lawrence Millman's* Goodbye, Ice, *a book that's equally an epitaph and a celebration for the arctic spirit-world and landscape. Said archeologists would say, 'So this is what happened here...'—and be haunted by it for the rest of their lives.*"
—Howard Norman, author of *The Ghost Clause*

"*Lawrence Millman is a polar bear of a man—explorer, ethnographer, mycologist. His poems take us into the Arctic wilds, introduce us to its icons, its relics, and its cultural curiosities. They bring you mementoes from the rapidly disappearing cultures of ice. When he prays, 'May the gods of the tundra grant me lichen until I become lichen myself,' take care. You may become lichen, too.*"
—Art Goodtimes, author of *Looking South to Lone Cone*

"*What Jacques Cousteau did for the oceans, Millman does for the Arctic, with the same sense of wonder and urgency. Finding beauty and humor in all he sees, he is a prophet in the wilderness. His gentle poems remind us a vengeful wrath awaits us if we don't repent. Never has a prophecy been so palatable. Let his flying shaman, his Inuit, raven, lemming, and bear take you on an exhilarating journey.*"
—David O. Born, author of *Eskimo Education and the Trauma of Social Change*

"In these poems Lawrence Millman lets us hear the voices of the North, of people who live across the frozen places telling their own stories and visions, people who accept they are one with all living creatures. Even the planet gets to have its say. The voices suggest that we pay attention, learn how to read our own geography so we can become one with our own landscapes."
—Claudia Radmore, editor of *Arctic Twilight: Leonard Budgell and Canada's Changing North*

"Millman's poems are playful, vulnerable, and magical—celebrating and grieving for our wild, animate, and chaotic world, and in the same breath bringing tundra-wide smiles, belly-deep laughter, and unbounded freedom."
—Michael Morrison

Goodbye, Ice

Arctic Poems

Also by Lawrence Millman

Our Like Will Not Be There Again
Hero Jesse
The Wrong Handed Man
A Kayak Full of Ghosts
Last Places
Wolverine Creates the World
An Evening Among Headhunters
Lost in the Arctic
Paris Was My Paramour
Northern Latitudes
Fascinating Fungi of New England
Hiking to Siberia
Giant Polypores & Stoned Reindeer
At the End of the World
The Book of Origins
Fungipedia

Goodbye, Ice
Arctic Poems

Lawrence Millman

COYOTE ARTS
Albuquerque, New Mexico

Goodbye, Ice: Arctic Poems. Copyright © 2020 Lawrence Millman.

Front cover photograph © 2015 Lawrence Millman, taken of Angmagssalik Fjord, East Greenland.

All rights reserved. No part of this publication may be reproduced, stored in a retrieval system or transmitted in any form or by any means, electronic, mechanical, photocopying, recording or otherwise without the prior permission of the publisher.

LCCN   2020938104
ISBN   Paper: 978-1-58775-031-1
       E-Book: 978-1-58775-032-8

10 9 8 7 6 5 4 3 2 1

Coyote Arts LLC
PO Box 6690
Albuquerque, New Mexico 87197-6690
www.coyote-arts.com

"When the well is dry, we know the worth of water."
— Benjamin Franklin

"Be in nothing so moderate as in the love of man."
— Robinson Jeffers

"Pissing in your shoe won't keep you warm for long."
— Icelandic Proverb

"What is the use of a house when you haven't got a tolerable planet to put it on?"
— Henry David Thoreau

Contents

Author's Note	xi
At the top of a cliff ...	15
Eider Duck	16
"Hey, look! A land bridge! ...	17
Aurora Borealis	18
There it rests ...	19
Raven	20
Seated in his canoe ...	21
Mosquitoes	22
A platter of seal liver and walrus fat ...	23
Goodbye, ice ...	24
Lichen	25
In the bright Arctic twilight ...	26
Kayaker's Song	27
Tupilaq Sonnet	28
Snowy Owl	29
Glacial Erratic	30
"Give me the moon ...	31
The concert begins ...	32
This is a poem ...	33

On this uplifted island ...	34
Lullaby	35
CLICK! ...	36
Sandhill Cranes	37
The bedridden Gwich'n elder ...	38
When our boat turns a bend ...	39
Never in the history of my eyes ...	40
Kelly Fraser, 1993–2019	41
All praise the mighty *Leccinum* ...	42
Jan Mayen Drift Logs	43
Icelandic Quatrains	44
There was a young girl named Isserfik ...	45
High above the shoreline ...	46
The Last Shaman	47
The Fall	48
The Methane Family	49
They've been grinning ...	50
Rust-ridden 55-gallon drums ...	51
I hear her now ...	52
Rising and descending ...	53
Civilization	54
I notice an all-white caribou ...	55
On the rough shingle ...	56
In my mind's grey asylum ...	57
Hunter's Song	58
There I was, bent over ...	59
What rests on the rock-ribbed short ...	60
Charm Against Disease	61
The wind huffs and puffs ...	62
There was once an Inuk ...	63

Night Music	64
A sled dog tethered to a post ...	65
As I paddled ...	66
Time-hardened arctic char ...	67
Charm Against an Avalanche	68
Invasive Species	69
There was once a family of skeletons ...	70
Berry Hunting Song	71
Up they rise ...	72
Inuit Elder	73
A sudden exaltation of wind ...	74
Walking the tundra ...	75
Marriage	76
A man dreamed of a good salmon fishing place ...	77
Ice Man	78
Squatting among tussocks of sedge grass ...	79
Epitaph	80

Biographical Note

Author's Note

Goodbye, Ice can be read as a poetical account of my wanderings in the Arctic between 1979 and 2019. Some of the poems are modified quotes from tradition-bound indigenous people in Greenland, Labrador, Arctic Canada, and Siberia; their words were written in prose in my notebook, but became poems in my memory. Other poems are horizontal rather than vertical—i.e., they're prose poems. A few are revisions of ones that appeared in *Northern Latitudes,* published in 2000 by New Rivers Press. Several poems are based on songs I collected in East Greenland in the 1980s.

In this book, I've tried not to revel in the self-absorbed ambiguities dear to so many contemporary poets, especially those who dwell in the halls of academe. Indeed, the poets I most admire are Walt Whitman, Robinson Jeffers, Sylvia Plath, Leonard Cohen, L. E. Sissman, Robert Service, Nazim Hikmet, and Vladimir Mayakovsky, none of whom ever dwelt in those not necessarily fertile halls.

As the book's title suggests, climate change appears not infrequently in its pages. This seemingly unstoppable phenomenon is among the reasons why I prefer snowy owls, lichens, ravens, bearded seals, musk-oxen, purple saxifrage, beetles, arctic poppies, sandhill cranes, and caribou to most members of a species once considered sapient—i.e., *Homo sapiens*. Likewise, I prefer Raven, the creator figure of the Chukchi, Koryak, Yupik, and Tlingit people, to a less beneficent deity named God (see the poem entitled "Raven" on page 20).

Lawrence Millman
Cambridge, MA
January 2020

Narsaq, West Greenland

At the top of a cliff
painted yellow by sunburst lichen
and directly above the blue-green
iceberg-dotted sea
with white-capped mountains
rising in the distance
rests a wayward heap of stones
an old Inuit burial cairn
with a jagged hole near the top
Gaze into this hole
and you'll see a mossy skull
minus a jawbone
and a few random ribs
along with ragged tufts
of sealskin leggings
stuck to an otherwise naked femur

Dear reader: you mustn't pity
this much-stripped person
for he or she has a beautiful view
for all eternity

Eider Duck

Krossbukta, Jan Mayen Land

On a frigid, wind-blasted day
with volcanic ash
coughed up by Beerenburg
coming down like a squall
she sits contentedly on her nest
and with the brownish-grey down
below her springy chest feathers
she gives her eggs a warmth
so exceptional
I can't help but say to her
with only a hint of anthromorphism
"Thanks, mother"

For Paul Kingsnorth

"Hey, look! A land bridge!"
shouted our Palaeolithic brethren
pointing at the Bering Strait
but just as they were starting out
on their celebrated journey
from Siberia to the New World
one of their shamans had a vision
of destroyed habitats
unrestrained development
permafrost methane emissions
and toxic waste everywhere
"Stop! We must go back!"
he cried out to his constituents
but his warning did no good
for humankind can only go forward
never back
never return to the less poisoned past

Aurora Borealis

Jade-green, flame-orange string figures spanning the night sky ... still-born children dancing with their afterbirths ... a celestial wolverine's neon jaws snapping open and shut ... the ghost fires of long-dead hunters cooking their prey ... spectral bears knocking around stars ... sky dwellers kicking walrus heads ... spumes of water ejected by aerial whales ... a giant concertina played by the great Moon Man

You are brighter, many times brighter, than the mundane sun

Wrangel Island, Siberia

There it rests
a relic of Soviet times
its floor a maze of moss
its walls collapsed
its seat not even a semi-circle
the northernmost outhouse in the world
Being in the *Guinness Book of Records*
or appearing on prime time TV
it doesn't give a damn about
for it has no interest in celebrity
It simply wants to return
and return as soon as possible
to the earth, its home

Raven

For Bernd Heinrich

As the brightest of all deities
Raven created the world
then he created its myriad denizens
bacteria, algae, protozoans
fish, plants, fungi, insects
avians, reptiles, mammals
an activity so tiresome
that he nodded off
whereupon a lesser deity named God
created a different type of organism
one that raped and pillaged
one that exterminated other species
one that even exterminated its own species

Raven awoke to an anthropogenic world
where God's organism reigned supreme
"Who made this two-legged monstrosity?"
he asked his grey-bearded inferior,
to which God proudly replied, "I did"
"Well, you can have it," Raven said
and then he flew off to create
a better world somewhere far, far away

Old Crow, Yukon Territory

Seated in his canoe
the Gwich'n man is crying
sputtering sobs of delight
His eyes are brimmed with tears
each one being a droplet of bliss
as he says, pointing skyward,
"The tundra swans have returned!"

Mosquitoes

Narsarssuaq, West Greenland

A falsetto whine
followed by the arrival
of flying hypodermic needles
reminds me
I'm part of the food chain
no different
than a bear, a caribou, or a bird
a fact that delights me
even as I crush
dozens of the little buggers

Upernavik, West Greenland

A platter of seal liver and walrus fat
with a side dish of berries
Nerillusuarisi! says my Inuit host
which means *bon appétit*
in his richly guttural tongue
I mispronounce the word
at least half a dozen times
so I ask him to write it down
but he shakes his head

Words are living things, he says
and we imprison them
by putting them on paper
just like we imprison a bird
by putting it in a cage

Poor words! I say to myself
even as I lock up a bunch of them
    in this cage

    Goodbye, ice
ancient membrane of the Arctic
goodbye, shimmering companion
who speaks in groans and cracks
roars, shrieks, and silences
you're melting into oblivion
and with your passing
the myriad lives you've blessed
polar bears, seals, and walrus
amphipods grazing on algae
and lipid-rich zooplankton
will become homeless
        forever more

Lichen

For Gail Coray

May the gods of the tundra grant me lichen until I become lichen myself. Let me be a rich yellow nitrophilous lichen decorating an Inuit burial cairn. Let me, a foliose lichen, stitch otherwise naked granite into a tapestry. Let me become a map lichen so I can read my own geography. May those gods grant me lichen, any sort of lichen, and I'll live on sunlight, dust motes, and the occasional drop of water, never complaining, never asking for more. Every hundred years I'll add an inch or two to my character. A millennium later, I'll still lack all presumption. And still grasp the rock of my choice with a full-bodied embrace.

On Baffin Island, Nunavut
For Bob Pyle

In the bright Arctic twilight
one Inuk scratches his head
another exclaims "Weird!"
and another shouts *"Kallupillak!"*
the Inuktitut word for evil spirit
and quickly jumps away
lest that spirit assault him
For none of them can fathom
this strangely colored
constantly flittering
otherworldly creature
a monarch butterfly
transported to these parts
by that evil spirit
known as climate change

Kayaker's Song

Heard in Upernavik, West Greenland

Damn you, kayak,
why are you going so slow?
Are you studying your lice
or maybe even fondling them?

Speed up, you sluggard,
for I need to get home
before this seal rots
before my wife rots, too

Tupilaq Sonnet

Tasiilaq, East Greenland

Once upon an earlier time
if you saw a hungry-looking skull
with two sharp beaks in lieu of ears
and even sharper mandibles
peering through your window
you'd be looking at a *tupilaq*
dispatched by the local shaman
to yank out your entrails

Nowadays if you see a *tupilaq*
you're looking at a souvenir
carved for Greenland's tourists
It won't yank out a tourist's entrails
but it will yank out cash from their wallets
Another victory for money over magic

Snowy Owl

Near Qamanituaq, Nunavut

A flesh-and-feathers gift of white
whose golden eyes flash like gemstones
flutters down onto a nearby boulder
and stares at me for a moment
then utters several low, rasping hoots
which inform me, "No offense, good sir,
but you're much less important
       than a lemming"

Glacial Erratic

For Elliott Merrick

All alone it stands, this headpiece of the world, far from the teeming fellowship of moraine, rubble, and till.

A palimpsest of *tripe de roche* gives a scurf-like skin to its robust body.

No other landmark graces this undulating tundra, twenty miles wide from eyelid to eyelid.

Its body is joined to the cold Labrador land by a harmony so strong that no one, not even Atikwapeo the Caribou God, could move it.

Pariah and bulwark, it offers a model of how to grasp the austere earth.

Grasp that earth, my friends, or you'll perish.

Told in Sermiligaq, East Greenland

"Give me the moon,
I want to play with the moon"
cried the little girl
So her parents visited the local shaman
a man who'd recently turned a missionary
into a muskox
but he refused to grab the moon
and give it to the little girl
He even refused to grab it
when the parents offered him a seal
even when they offered him two seals
For what's in the sky
should remain in the sky,
just as what's on the earth
should remain on the earth

Angmagssalik, East Greenland
For Helmut Schöner

The concert begins
one dog howls a falsetto moan
another follows with an alto voice
another with a cavernous bass
yet another with a high-pitched soprano
and soon the whole village is howling
some dogs stick to the same key
others change keys
some howl plangent octaves
others rip the air with sforzandos
some seem to be howling Beethoven
others Bela Bartok
still others oldies but goodies
and a few even Karlheinz Stockhausen
yet each dog's howl
whatever the key or musical style
carries the same message:
O human master,
let me pull a sled, please,
or I'll feel like a pekingese

Herschel Island, Yukon Territory

This is a poem
that can't be taught
in a walled-off classroom

After all, it doesn't boast
a single ambiguity
for a teacher to disentangle

Its meaning is just as clear
as the sunlit terrain
of this high latitude island

Do not bow to religion
or the raging hyenas
of capitalism

Bow only to rocks
bow low to them
for they know how to live

Kolyuchin Island, Chukotka

On this uplifted island, the ground is a lovely blanket of purple Chukchi primroses *(Primula tschuktschorum)* except where it's an unlovely blanket of debris from an abandoned Polar Research Station from Soviet times.

The debris consists of the following items: the remains of a funicular railway; a shattered weather vane; a large anemometer; 55 gallon oil drums; and several badly rusted tins of cat food.

I pick up a chunk of matting that had once been a book. "Pushkin," my Russian companion says. She gently fingers the matting, then says, "Poor Pushkin." There are tears in her eyes.

Lullaby

Heard in Tasiilaq, East Greenland

Go to sleep, my little one
go to sleep
or else a raven will fly down
and peck out your eyes
and *then* you'll go to sleep
my dear sweet little one

CLICK!
> CLICK!
>> CLICK!
goes the photo feeding frenzy
by cruise ship passengers
taking endless pictures
of a huddle of walruses
reclining on an Arctic beach
The walruses can see the passengers
bipeds wrapped in orange parkas
but has a single one of those bipeds
with cellphones and high end cameras
blotting out their eyes
actually seen a walrus?

Sandhill Cranes

Near Selby Lake, Alaska

    Gliding south
a triangular congregation
of outstretched necks
and echoing bugle calls
circles above my head
not once or twice, but three times
making me feel more blessed
than if a host of winsome angels
gracefully flying across the sky
    anointed me

Old Crow, Yukon Territory

The bedridden Gwich'n elder isn't worried about dying. After all, everyone dies sooner or later, he observes, so how can you disapprove of such a popular pastime? No, what worries him is this: if he's dead, he won't be able to eat anymore ... and eating is the most pleasurable activity in the entire world. The Afterlife may have a virtue or two, he says, but you can't even nibble on a piece of pemmican there. Whereupon he grabs a chunk of caribou flank and happily stuffs it in his mouth.

Porcupine River, Yukon
For Paul Shepard

When our boat turns a bend
I see the tusk of a woolly mammoth
a creamy-white artifact
from the late Pleistocene
emerging from an eroded cutback
Is it my imagination
or could that tusk be pointing
at present-day passersby
in canoes and motorboats
and asking this question:
How could paltry creatures like you
have survived
and I did not?

Near Cambridge Bay, Nunavut

Never in the history of my eyes
have I witnessed a monster
like this rough beast
slouching not toward Bethlehem
but across Victoria Strait
and releasing noxious gases
into the previously pristine air
as well as dumping tons of raw sewage
into the previously pristine water
O the horror! the horror!
What kind of beast is this?
It's a 13 deck cruise ship
called the Crystal Serenity
traversing a Northwest Passage
now so ice-free
any monster can traverse it

Kelly Fraser, 1993–2019

How well I remember
her dancing eyes
and the unencumbered smile
on her eight year old face
as she sliced seal meat for me
under Sanikiluaq's flaming aurora

She became a popular Inuit singer
so popular she took drugs
and so drug-addled
she killed herself
eighteen years later
far from any flaming aurora

Baker Lake, Nunavut

All praise the mighty *Leccinum,*
a scaber-stalked mushroom
that readily spreads its spores
in a habitat so wind-blown
and likewise so frigid
that if a redwood,
seemingly mighty itself,
somehow found itself here,
it might bawl these words,
"California, where art thou?"

Jan Mayen Drift Logs

For Bob Blanchette

From the sea's catacombs come these dearticulated bones, shuttled by Arctic currents from Siberian rivers to Jan Mayen's distant treeless shores. Each bone is a log shellacked to a primordial smoothness by the frigid fingers of northern seas. The most recent are a chestnut brown, as yet undone by the high latitude sun, while the oldest have turned a friable grey. And on grey shingle they lie, askew, isolate, or piled high, a graceful sculptural dream to those of us who are just passing by.

Icelandic Quatrains

For Vilborg Dagbjartsdóttir

Learning
Higher education comes to Arnanupur
In the derelict schoolhouse
Elias of Sveinseyri
hangs his shark's meat

Shipwreck
The *Baldur* strikes at Eldey
Its dead crew wash ashore
wrapped in seaweed
a cheaper garment than a shroud

Northern Lights
A giant concertina
shimmers across the sky —
why bring any fiddles
to tonight's dance?

Puffins
Tuxedo-clad
the avian politicians
inspect their ocean empire
and end up in the pot

Sung in Tinit, East Greenland

There was a young girl named Isserfik
O she was a fine piece of blubber
and any man who saw her loved her
O Isserfik, why don't you fancy me?
Because I fancy eagles more, she'd say
Give me an eagle over a man any day
Now there was an eagle from Qassiasut
He carried Isserfik off to his nest
and in that nest together they lay
Nestled together happily they lay

O Isserfik, what's that in your *amaut* sack?
It's the beautiful little girl I've hatched
    half human, half eagle
    and dearly do I love her

Novoye Chaplino, Chukotka

High above the shoreline
a giant whale carcass
year in and year out
offers the air its smell

This cornucopia of rottenness
doesn't bother
the local Yupik and Chukchi
who live with the dead whale

After all, decay is part of life,
they'll tell you,
so why hold your nose
when nature wants in?

The Last Shaman

Angmagssalik, East Greenland

Bedridden he is
this bundle of age
who once could fly
fly high in the sky
just by flexing his index fingers

Songless he is,
this man of songs,
who once could chant away
storms and wild winds
with the guttural of his voice

And full of sickness he is,
this ancient healer,
who once could cure anything
from festering wounds
to possession by evil spirits

Now there's nobody left
to cure him
Flying is easy,
he seems to say,
it's the not flying that's hard

The Fall

Tombstone Park, Yukon Territory

When I entered the Garden of Eden
a green blanket of caribou lichen
punctuated by tufts of arctic cotton
I soon saw (not an apple!)
but a sumptuous bakeapple
a both sweet and bitter berry
so dear to my heart
that whenever I see it
I feel red's the color of my true love
So I reached for this bakeapple
tripped and fell down
but unlike those poor sods
our perpetually fallen First Parents
I got up again
and grabbed the bakeapple
The absence of God smiled

The Methane Family

The northern permafrost is melting
and out totters Mr. Methane
with a quizzical look on his face
Where do I live now? he wonders
He scratches his head
then scratches it again
and comes up with the answer
That answer is indeed "up"
So flexing the myriad muscles
you might expect
from such a potent greenhouse gas
and flapping his newborn wings
he rises up, up, and up
to carouse in the atmosphere
with his fellow greenhouse gases
Whereupon the permafrost melts even more
and out totters Mrs. Methane with her kids

Qajartalik Island, Nunavik

They've been grinning
for at least a millennium
these granite-dwelling petroglyphs
with faces of shamans
or perhaps just of cheery folks

But one of them isn't grinning
It has a vehement cross
carved directly onto its face
and it's now gazing at the world
not with a grin but a grimace

Wrangel Island, Siberia

Rust-ridden 55-gallon drums
survivors from the Soviet era
stand like sentries
all along the rocky shoreline
Once they were full of oil
but now they're full of a hauteur
that informs us mere mortals
"Come back forever
and we'll still be here"

Northwest River, Labrador

I hear her now
a delicate little brook
giggling at me
she's so bashful
she hides under the snow
but she still giggles
sometimes even chuckles
and her sweet voice
dismisses the cold

Grimsey, Iceland
For Lene Zachariessen

Rising and descending
soaring and dropping
the Arctic terns scream *Kria! Kria! Kria!*
as they dive-bomb me

They're protecting their eggs
or so it's commonly assumed
but they perform their winnowing aerobatics
glissandos and pirouettes
bends, bows, and whiparounds
with such elegance
I can't help but think of them
as a *femme corps de ballet*
who've emancipated themselves
from the ordinary stage
and whose eggs matter less to them
than their elegant movements

And so when their beaks peck my head
I think of my blood
as payment for their choreography

Civilization

From Rigolet, Labrador, to London
traveled Caubvick
a pretty young Inuit woman
This is civilization, my dear,
the English informed her
She returned to Labrador
no longer pretty
for she had acquired smallpox
there were pustules on her face
and most of her hair was gone
So is it any wonder
that her fellow Inuit
to whom she bequeathed smallpox
now regarded civilization
as a disease?

Baker Lake, Nunavut

I notice an all-white caribou munching on a field of *Cladonia* lichen. "Spirit animal," observes my Inuit companion, adding, "and if you take a photo of it, you will be buried forever by a snowstorm."

I now gaze reverently at the caribou, which gazes back at me, then produces a bountiful outpouring of fecal matter — a reminder that even sacred creatures are obliged to answer nature's call.

Utshimassits, Labrador

On the rough shingle
a canoe washes up
It seems to be empty
but if you look closely
you'll see an Innu hunter
lying in the bottom of the canoe
his head crushed
chest excavated
and one thigh missing
Red twists and zigzags
the claw marks of a bear
etch the man's remaining flesh
and carry this boast:
I'm a better hunter than you

Baker Lake, Nunavut

In my mind's grey asylum
I rummage for something
as brightly yellow
as this solitary arctic poppy
rising from the tundra

At last I have the answer
it's the bright yellow arctic poppy
I saw yesterday afternoon
now lighting up another chamber
in my mind's grey asylum

Hunter's Song

Heard in Kuujarut, Nunavut

Dear caribou
sweet caribou
don't be shy
come and visit me
you'll be so pleased
because my wife
my dear sweet wife
she'll be chewing your skin
to make *kamiks* and mittens
such nice *kamiks* and mittens
you'll like them very much

Wrangel Island, Siberia

There I was, bent over
a heap of polar bear shit
studying its contents
Here were seal whiskers
here were berry pits
and here was a delicate maze
of bird bones
Suddenly I looked up
and there was a polar bear
studying me
You're an advanced species
and *this* is what you do?
its baffled expression seemed to say
Then it loped away
rejecting me
before I could be afraid

Camsell Island, Nunavut

What rests on the rock-ribbed shore
of this uninhabited island
a place not on any map
(true places seldom are)
what but
        a sodden
                box of Pampers
which just goes to show
that even at the end of the world
there's no escaping human wipes

Charm Against Disease

Sung in Ikateq, East Greenland

White gull
who soars over my dismal body
I'm talking to you, bird
Come down from the sky
and grasp me with your wings
Let me fly in the blue air, too

Near Okak, Labrador

The wind huffs and puffs
it bellows and roars
shouts and shrieks
then huffs and puffs again
but it doesn't blow a house down
After all, there aren't any houses here
only boulders
and if you listen closely
very closely
you might hear one of them say,
"How about another kiss, dear wind?"

Heard in Repulse Bay, Nunavut

There was once an Inuk
who had no friends
not even a single one
One day he went for a walk
and he sank in some moss
sank deeper and deeper
until he disappeared
I'm comfortable down here, he said
and he remained in the moss
until he became moss himself
Now he had friends all around him

Night Music

Near Utshimassits, Labrador

Here I am, answering nature's proverbial call on a night so cold that spruce branches crack, then break off and fall with muffled thuds in the snow. The cold skitters along my bones like rambunctious fingers on a keyboard. Overhead, the stars are kicked up snowflakes, glistening pinpricks decorating the blue-green streams of the aurora. I gaze at them until the wind sings these words to my nearly numb flesh: now that you've answered your call, it's retreat or die. So back I go, half-frozen yet fully-blessed, back to my tent's cozy life-giving unloveliness.

Kulusuk, East Greenland

A sled dog tethered to a post. A raven hops onto its back, and the dog swivels its head repeatedly, trying to bite the intruder. *Kra-a-a-a, turna-r-r-rk, gga-agga, ker-r-r-rk!* proclaims the raven. Translation: You've not very good at this, are you? Meanwhile, another raven has gobbled up half the dog's fish dinner.

Now the fully-fed raven jumps onto the dog's back, inspiring more failed bites, while the other one finishes the fish dinner. They fly off together, but not before they give the dog a patronizing look, then say, *Kar-r-r-rk, kra-a-a-kr, ga-ga-rk!* Translation: It's a shame you allowed yourself to be domesticated, old chap.

Near Kanjiqsujuaq, Nunavik

  As I paddled
the blue waters of Hudson Strait
the last thing I expected to see
was *(mirabile dictu!)* an iceberg
especially a winged iceberg
with a multiplicity of turrets
rising from its buried body
I was gazing at it in wonder
when it suddenly calved
and an enormous wave
came hurtling in my direction.
"Here's my gift to you, Humanity,
for melting the sea-ice, my kin,"
the iceberg seemed to be telling me
I turned my kayak around
and quickly paddled away
for I had no wish to receive
  my just reward

Igloolik, Nunavut
For Georgia

Time-hardened arctic char
with honeycombed bodies
filled with sand and pebbles
the flotsam bones of birds
their salt-widened eyeholes
gaping at the wide wide world
a cream-colored whale's vertebra
broken whelk and limpet shells
wrapped in knots of seaweed
populate this shingle shore
with the delicacy of dismemberment

Charm Against an Avalanche

Heard in Kummiut, East Greenland

You mighty heap of rocks
I'm just a weak little human being
who cowers at your feet
If you wish to flex your muscles
bury a cliffside or two
but, please, I ask you
please don't bury
a worthless thing like me

Invasive Species

Clad in the paper-thin garb
of their sun-dried homeland
and craving *baba ganoush* and *falafel*
God and his son huddle together
in their high end polyester cotton tent
"It's friggin' cold!" God declares
A shivering Jesus nods in agreement

They'd managed just two conversions:
a mentally challenged lemming
and a recently deceased caribou
"Screw this place!" proclaims God
to which Jesus says, "Amen to that, dad!"
So they hop onto their private jet
and fly back to the Holy Land

Whereupon there's a sudden gust of wind
the Arctic sighing with relief

Heard in Upernavik, West Greenland

There was once a family of skeletons living in a little hut — a grandmother skeleton, a father and mother skeleton, and two little boy skeletons. They managed just fine because they didn't need anything to eat. One day a hunter arrived and stayed in their hut. The skeletons emerged from the ground and began to sing and beat their drums. The hunter burst into laughter. "You are only skeletons," he told them, "so stop acting as if you were human beings." The skeletons were so ashamed that they retreated into the ground and never beat their drums again.

Berry Hunting Song

Heard in Sermiligaq, East Greenland

Where can I find berries?
Who can I ask?
Not a raven
nor a reindeer
and certainly not a polar bear
but a little lemming
a smart little lemming

Just follow me, he'll tell you,
we'll go up that hillside
and I'll show you berries
all the berries you want
berries so dew-fresh
and so sweet
only a lemming can find them

Chukotka, Siberia

Up they rise
corpses from the Kolyma gulag
wearing the same prison garb
they were buried in

They were glad to be dead
since no guard ever whipped them
They were never hungry
Never frostbitten, either

Now their deaths have been violated
by dredges and wheel bucket excavators
savaging their former resting place
in search of gold, tin, and tungsten

So away they totter
hoping to find unbroken ground
into which they can bestow
their last, their very last remains

Inuit Elder

    Only a man like him
who still believes in birds and beasts
    and not foreign deities
can know that when he pisses skywards
    the long yellow stream
will travel all the way to the moon
    and illuminate it

Selby Lake, Alaska
For Art Mortvedt

A sudden exaltation of wind pushes across the lake, inspiring hungry waves to rush the gunwales of my canoe. Each paddle stroke thrusts me backwards, farther and farther from shore, where stunted spruce now gesticulate like semaphores gone berserk. My arms are a pair of broken matchsticks, and my face is incised by needles of cold spray. There's no way I can resist such an elemental force, so I lay down my paddle and let the wind caress me with fingers dipped in the Arctic Ocean.

Near Ferguson Lake, Nunavut
In memory of Pentti Linkola

Walking the tundra
I don't need a GPS
for I can easily see
purple saxifrage and blue harebells
sedge meadows and fell fields
reindeer lichen and willow thickets
arctic cotton and lemming burrows
glacial erratics and glacial rubble
all of which tell me where I am
and, indeed, who I am

Marriage

Give me the earth's scoured bones,
says *Diapensia lapponica,*
a perennial evergreen shrub
that scorns grassy places
preferring rocky ledges
and barren scree slopes
as well as glacial till

Other plants drop their dead leaves
but the savvy *D. lapponica*
wears its dead leaves
like a down parka
so that frigid winter after frigid winter
it can still marry
the earth's scoured bones

Heard in Sheshatshui, Labrador

A man dreamed of a good salmon fishing place by a river. He went there, sat down on the bank, and fished for one whole month. All he caught was a cold in his chest. I'll kill that damn dream, he said to himself. So he went to sleep holding a knife in his hand. The dream told him of another good fishing place. The man thrust his knife, then thrust it again. The dream let out a soft sigh and died. Now the man didn't even know where the bad fishing places were, and he died, too.

Ice Man

For Noah Piugaatuk

An Inuk at a seal's breathing hole
his harpoon ready to strike
a figure of infinite patience
airplanes rip across the sky
satellites orbit the heavens
but the Inuk doesn't look up
does not even move his eyelashes
he remains gazing at the hole
waiting, waiting, waiting
for the seal to raise its head
one final time
before the ice disappears

Herschel Island, Yukon Territory

Squatting among tussocks of sedge grass rest the dead boards of a church — collapsed windows, permafrost floor, a few lupines at the door, and a fruiting of *Russula* mushrooms in the nave. The nearest thing to a crucifix is a rotting pair of crossed boards fallen from the roof.

Yet what better place for divine worship? In the decayed rafters sits the congregation — a group of black guillemots. They proclaim their faith by piping high-pitched hymns of praise to the Arctic's gaping solitudes. As I listened to them, I too became devout.

Epitaph

For John Hanson Mitchell

Please help me,
I need to be cold
or I'll die,
cries the Arctic

Sorry, old chap,
but it's much more fun
to be hot,
says *Homo sapiens*

happily reclining
on a beach chair
even as the rising seas
begin licking at that chair

Biographical Note

Writer-ethnographer-mycologist Lawrence Millman has made over 40 trips and expeditions to the Arctic and Subarctic. His 18 books include such titles as *Last Places, Northern Latitudes, A Kayak Full of Ghosts, Our Like Will Not Be There Again, Hiking to Siberia, Lost in the Arctic, At the End of the World, The Book of Origins,* and *Fungipedia.* He has written for *Smithsonian, National Geographic, Outside, Atlantic Monthly,* and *The Sunday Times* (London). He lives in Cambridge, Massachusetts.

www.ingramcontent.com/pod-product-compliance
Lightning Source LLC
Chambersburg PA
CBHW050042080526
44586CB00014B/1412